I0494616

Designers' Essentials Series

INSULATION: CREEPAGE, CLEARANCE AND SOLID INSULATION

AN INTRODUCTION BY PROFESSIONALS

Basic theory and numerical calculation examples on the most important topic of electrical and electronics design.

MARCO CATANOSSI

Summary

Dedication

To D.

Introduction

This book is intended to be a demystifying introduction to particular aspects of electrical and electronics design concerning insulation: clearance, creepage and solid insulation (or isolation, the terms have the same meaning here).

The human body is modelled as hand to hand 1 KΩ resistor in dry conditions.

This implies that when a voltage is applied to it, e.g. touching a live part, a current will flow.

This current produces harmful effects only if it exceeds a minimum value. At 0.5 mA 50-60 Hz it can cause a startling reaction on our hands. At higher values we may experience the inability to let an arm go (35 mA, 15-100 Hz) and at 10 mA 15-100 Hz along a hand to foot path, there is the possibility of ventricular fibrillation and cardiac arrest.

In everyday life, electrical products also have currents flowing along unwanted paths, the so called "leakage currents", that when we touch external metal parts (e.g. a washing machine) it might discharge to earth through our body.

In order to keep these currents under a safety value, the insulation must have a minimum resistance, usually in the MΩs.

The objective of all the electrical safety measures is to avoid both electric shock and that the leakage currents exceed certain safety values.

In this book, first we will explain basic concepts and definitions, then we will proceed with practical calculation examples of clearance and creepage distances for double and reinforced isolation taking as an example a food mixer appliance design.

We will also talk about solid insulation in a specific chapter, with an example on how to choose a solid barrier or enclosure.

Finally we will introduce insulation diagrams, a simple and powerful representation tool that is often neglected.

The state of the art on the safety techniques can be found in international standards, which are continuously updated and are agreed among the National Agencies, industries and other stakeholders.

Standards are used worldwide by governments for regulatory purposes and generally to set up a bottom line for safety requirements, so we will refer to them throughout the book.

The Author is a professional involved in everyday works on the matter and has used a double approach:

• Theoretical definitions (being kept to what essential is) and

• A practical example to explain calculation procedures

with the paramount objective to keep it short and practical.

Chapter 1 – Definitions and Basic Concepts

1.1 General Safety Requirement Clause

Almost every standard dealing with product safety includes a general safety requirement clause that requires that the product must not generate hazards not only when properly used or well-functioning, but also under expectable carelessness and misuse.

This requires particular attention from the manufacturer/designer, who can be considered liable also when the product has been misused along expectable patterns, such as:

- lack of concentration or carelessness

- choosing the line of "minimum resistance" when performing an action

- common reflex in case of malfunction, incident, failure, unexpected event

Furthermore, a product must be designed so that is still safe under single fault conditions. This approach has its rationale in the concept of "double improbability", the probability that both the safety measures fail is so low that it can be considered acceptable.

This rationale underpins the electrical safety measure called *double insulation* that protects us against electric shock resorting to:

- a *basic insulation*
  and

- a *supplementary insulation,*

involving in this way two means of protection.

Nonetheless, we may still accept just a means of protection, if this means has such safety features to make it equivalent to double protection. In this case, we talk about *reinforced insulation.*

At the end of the day, insulation is about clearance, creepage and solid insulation, which require to calculate in advance distances and choose the proper materials.

1.2 Product Classification

Before we leave this chapter and start talking in details about these insulation measures, it is useful to add some more concepts and classifications.

With respect to electrical insulation and safety product classification, we have:

Class I Products: products in which protection against electric shock relies on:

- basic insulation

and

- the connection of conductive accessible parts to the protective earthing conductor, avoiding such accessible parts to become live in the event of a failure of the basic insulation.

It relies on the good quality of the building earth installation. It has no particular symbol on the label that identifies it. In order to recognize if a product has a Class I protection against electric shock, we may assume that if no other symbols are present (see class II or III), the product is in class I.

Class II Products: protection against electric shock relies **on**

- double insulation (basic + supplementary insulation) or
- reinforced insulation.

There is no protective earthing or reliance upon installation conditions, so the product is inherently safe by design. A product in class II carries this symbol on the label.

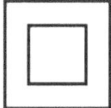

Class III Products: this solution relies upon SAFETY EXTRA-LOW VOLTAGE (SELV, less than 42 V DC or peak between conductors and between conductor and earth). Inside the product there are no generated voltages higher than 42V. If it is in class III, the product label carries this symbol.

Chapter 2 – Clearance

2.1 Definition

When designing the insulation of an application, basically we need to take care of distances. These distances are:

• Air distance from live conductors

• Distances along a surface

• Thickness of a barrier

These distances need to be chosen so that no breakdown happens, which can hurt the user or cause any danger.

The air distance between/from live conductors is called clearance distance.

Clearance is the shortest distance in air between two conductive parts or between a conductive part and a surface, which is considered accessible by the user (e.g. an external surface, or an internal conductor accessible by using a finger through an opening in the enclosure).

This spacing is filled with air and a spark-over could happen if there is an electrical breakdown.

As an image, we may think of the distance covered by a fly from one place to another.

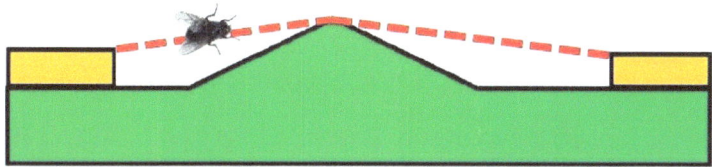

Figure 1 - Clearance distance

2.2 Factors Influencing Clearance Distance

This spacing needs to be higher than a minimum safety distance and depends on:

- **The insulation working voltage**: for higher voltage, we need more clearance spacing.

  If the working voltage (the voltage inside the product) is higher than the rated voltage (the supply voltage), there are often further consideration and reference to particular standards, like the EN 60664-1. In these cases, the clearance is assumed to be the highest value of different calculation options.

- **Type of circuit:** either primary (powered by the main) or secondary (after a transformer).

  As a general rule, the primary circuit (the one connected to the wall socket) needs more protection (i.e. more clearance) than the secondary one.

  The secondary circuit is protected by a separation device that adds safety and filtering. This aspect will be

11

reflected in the overvoltage category, which directly influences the clearance distance (see later).

- **Pollution degree:** if the air is polluted or if the presence of moisture can be foreseen at the intended use location, the clearance needs to be higher.
  This is because the pollution and moisture usually include a conductive factor that facilitates a spark-over.

  Examples:

  - Pollution degree 1: Clean room environment, within sealed components.

  - Pollution degree 2: IT equipment, laboratories equipment, test station, office environment.

  - Pollution degree 3: electrical equipment in industrial and farming areas, unheated rooms, boiler rooms, kitchen machines.

  - Pollution degree 4: electrical equipment for outdoor use.

- **The insulation category that is meant to be implemented:**
  e.g. basic, supplementary, reinforced.
  The standard calculation procedure that we are going to explore, suggests minimum distance values that generally

differ when we design a basic, supplementary or reinforced insulation.

- **Altitude:** at a higher altitude, there is less pressure and we need a higher clearance.

  Altitude is considered to be a sort of multiplication factor in the calculation procedure.

- **Overvoltage category:** depending on how far the application is from the supplying network, we have more or less overvoltage to withstand.

  The overvoltage category takes the electrical cleanliness of the supply voltage into account. The supply line at a distribution level is "dirtier" and we may expect a higher disturbance peak than after a filtered wall outlet. Therefore, the user needs more protection and clearance distance.

- **Quality Control Program:** for some specific applications, for instance if we design an Information Technology equipment, if we have a quality control program in place, this could enable a thinner clearance than otherwise imposed. This is not generally true for other kinds of equipment, like for example in case of household appliances, where such a benefit is not granted.

2.3 Overvoltage Categories

As far as overvoltage categories are concerned, applications fall into one of the following categories:

Category I - For connection to circuits in which measures are taken to limit transient over-voltages to an appropriately low level.

Examples: Protected electronic circuits, secondary circuits, USB powered peripherals (where the computer power supply transformer has already cleaned the line).

Category II - Energy-consuming equipment to be supplied from the fixed installation.

Examples: Appliances, portable tools, and other household appliances and similar loads.

Category III - In fixed installations and for cases where the reliability and the availability of the equipment is subject to special requirements.

Examples: Switches in fixed installation and equipment for industrial use with permanent connection to the fixed installation, indoor wiring.

Category IV - Used at the origin of the installation.

Examples: Electricity meters and primary overcurrent protection equipment (distribution line transformer).

Figure 2 shows where, from a conceptual level, the four overvoltage categories are nested.

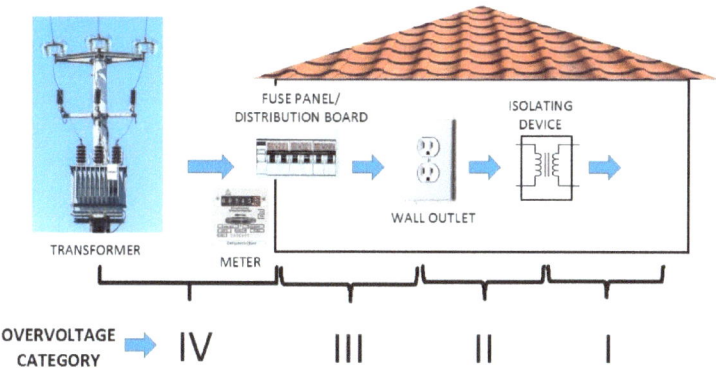

Figure 2 - Overvoltage categories

2.4 Calculation Procedure for a Food Mixer

Let us continue with a practical example. As explained above, before designing a circuit, we need to define the following points:

- Insulation working voltage

- Type of circuit, whether it is primary or secondary

- Pollution degree (1 to 4)

- Insulation category to implement (e.g. basic, supplementary, reinforced)

- Altitude

- Overvoltage category (I to IV)

15

- If we have a quality control program in place

We're going to design a household appliance, and specifically a food mixer, plugged to the wall socket and marketed in Europe (important to know, because of the differences in network voltages with the U.S.).

We profile our application as follows:

- Insulation working voltage --> 220 V, no internal higher voltage

- Type of circuit, if it is primary or secondary --> only primary

- Pollution degree (1 to 4) --> pollution degree 3

- We are going to consider the insulation category that is meant to be implemented (e.g. basic, supplementary or reinforced)

- Max altitude of use --> less than 2000 metres

- Overvoltage category (I to IV) --> it is plugged into the wall outlet, so it is II

- If we have a quality control program -->, it does not make any difference for household appliances (as for IEC 60335-1)

First, we find the applicable standard, that in this case, for household appliances is:

- IEC 60335-1 for Household and similar electrical appliances, and

- IEC 60335-2-14 for electric kitchen machines with rated voltage being not higher than 250 V.

Then we will proceed according to the diagram in figure 3.

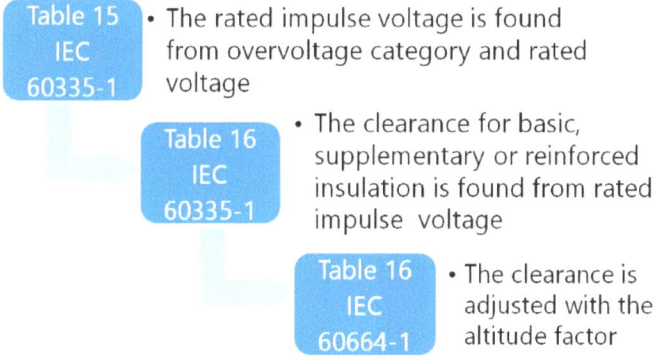

Figure 3 - Calculation diagram

2.5 Numerical Calculation Example: Basic Insulation

From our rated voltage (220 V) and overvoltage category (II), we are able to find our rated impulse voltage, in this case 2500 V, according to Table 15 IEC 60335-1.

Rated voltage V	Rated impulse voltage V Overvoltage categories		
	I	II	III
V <=50	330	500	800
50 < V <= 150	800	1500	2500
150 < V <= 300	1500	2500	4000

Figure 4 - Table 15 of IEC 60335-1

With this value, we will refer to Table 16 for the minimum clearance value that we need to adopt.

Rated impulse voltage V	Minimum clearance mm
330	0,5 (1) (2)
500	0,5 (1) (2)
800	0,5 (1) (2)
1.500	0,5 (1)
2.500	1,5
4.000	3
6.000	5,5
8.000	8
10.000	11

Figure 5 - Table 16 of IEC 60335-1

The clearance marked with (1) is increased to 0.8 mm for pollution degree 3.

Furthermore, for PCB tracks the clearance marked with (2) is reduced to 0.2 mm for pollution degree 1 and 2 .

From Table 16, our minimum clearance space is 1.5 mm. The pollution degree is 3 (imposed to be considered so by IEC 60335-

2-14), but in our case the value found does not need any correction, according to the notes in Figure 5.

Now it is time to consider the altitude of use.

The standard IEC 60335-1 refers to another standard, IEC 60664-1. The latter recommends the following correction factor for clearance.

Altitude m	Multiplication factor for clearance
2.000	1,00
3.000	1,14
4.000	1,29
5.000	1,48
6.000	1,70
7.000	1,95
8.000	2,25
9.000	2,62
10.000	3,02
15.000	6,67
20.000	14,5

Figure 6- Table 16 of IEC 60664-1

So in this case no multiplication is needed, because the maximum altitude does not exceed 2000 metres.

This means that the mixer is not allowed to be used at Lhasa, Tibet 3650 meters, or Leadville, Colorado 3094 meters. This altitude limitation must be well written on labels and user manual, together with its advertising material.

Until now, we have found what distance we need between live conductors or between live parts and accessible surfaces in order to have a basic insulation: 1.5 mm clearance makes us implement this first stage of protection.

If we design a Class I product, these calculations are enough, because the second protection measure is the connection to protective earth. For a Class II products instead, we must continue with the supplementary insulation.

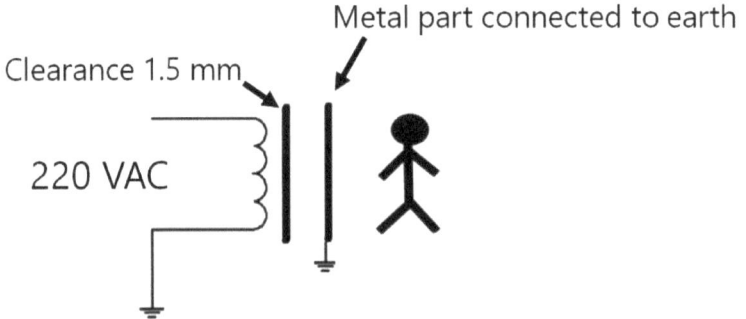

Figure 7 - Class I device

2.6 Numerical Calculation Example: Supplementary Insulation

According to IEC 60335-1 Standard, the clearance for the supplementary insulation cannot be less than that specified in Table 16.

Therefore, if we design a Class II product, with a double insulation, our clearance distance for the supplementary one is at least a further 1.5 mm.

We can conclude our clearance distance design with a 3 mm clearance distance, which ensures double insulation against breakdowns through airgaps.

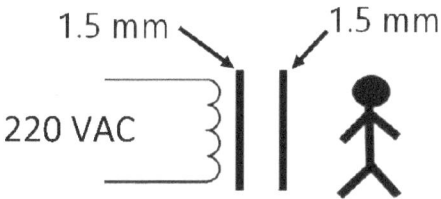

Figure 8 - Class II device

2.7 Numerical Calculation Example: Reinforced Insulation

On the other hand, if our isolation strategy is to provide double safety with a reinforced means of protection, the standard IEC 60335-1 states that for reinforced insulation, we still have to refer to Table 16, but using the next higher step for rated impulse voltage as a reference. So, instead of the clearance corresponding to rated impulse voltage 2500V, we will refer to the following value corresponding to 4000V.

Rated impulse voltage V	Minimum clearance mm
330	0,5 [1] [2]
500	0,5 [1] [2]
800	0,5 [1] [2]
1.500	0,5 [1]
2.500	1,5
4.000	3
6.000	5,5
8.000	8
10.000	11

Figure 9 - Table 16 of IEC 60335-1 for reinforced insulation

With a 3 mm clearance, we have also implemented a reinforced insulation (please notice that the coincidence of this value with the double insulation is fortuitous, generally this may differ).

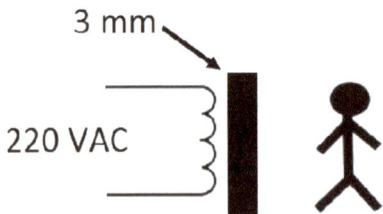

Figure 10 - Class II, with reinforced insulation

2.8 Others Fields of Application

We have taken the household standard IEC 60335 series as an example.

If our product falls into other categories, for instance Information Technology, then the IEC 60950-1 is applied and our calculations

22

bring to a clearance of 2 mm that could be reduced to 1,5 mm if the manufacturer has a Quality Control Program in place.

On the other hand, if we design a medical device, IEC 60601-1 is applied and the clearance value may change accordingly, for instance, if the exposed person is an operator or a patient.

In conclusion, every standard has some specific requirements that may change, as the standard updates over time.

So before starting the design process, it is important to:

1- Find the relevant standards available for the specific application.

2- Purchase the latest version of the standard.

3- Study it carefully.

Chapter 3 – Creepage

3.1 Definition

Together with clearance, we must also take care of the Creepage Distance.

The creepage distance is still measured between two conductive parts or between a conductive part and an accessible surface. But this time, instead of "in the air", we have to travel along the surface of an insulation material. As a graphical example, think of the path a snail would follow.

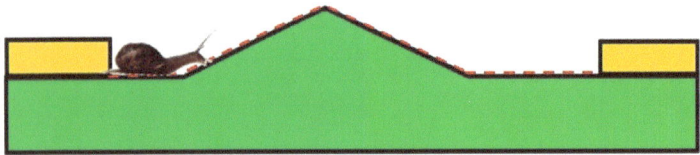

Figure 11 - Creepage distance

Creepage is a parameter that must be kept in mind in particular during PCB Design.

If we do not have enough creepage, we can change the physical geometry to match the requirements.

For instance, digging a space between traces.

Figure 12 - Creepage enhancement

Or the addition of an insulation barrier.

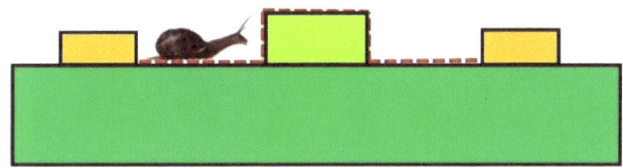

Figure 13 - Creepage enhancement

Creepage also affects distance between:

- tracks and conductive parts nearby like screw through mounting holes, and
- distances between primary and secondary circuits, for instance when the secondary circuit exchanges some form of energy with a patient body during a medical device application.

Red dashed lines in Figure 14 show these cases.

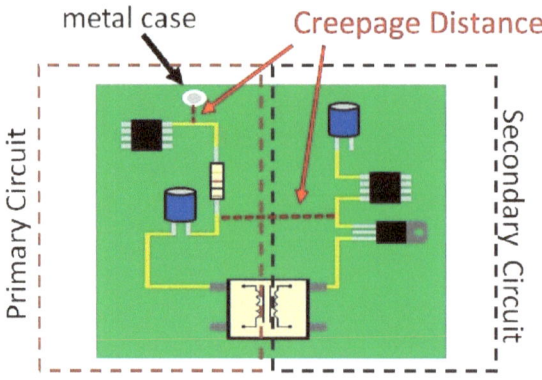

Figure 14 - Creepage on a PCB

3.2 Factors Influencing Creepage distance

When we design a creepage we have to consider several parameters, with some differences with respect to clearance. Here we can see what applies:

- Insulation working voltage --> **YES**

- Whether the circuit is s primary or secondary --> **YES**

- Pollution degree (1 to 4) --> **YES**

- Insulation category (e.g. basic, supplementary, reinforced) --> **YES**

- Altitude --> **NOT ANYMORE**, usually the insulation material does not change its characteristics with altitude

- Overvoltage category (I to IV) --> **YES**

- If we have a quality control program --> **NOT ANYMORE**

Furthermore we have to take care of the kind of surface material:

- The material group category (I, II, IIIa, IIIb). The index used is the Comparative Tracking Index, CTI (per IEC 60112) --> **NEW**

When we apply a voltage difference to two conductive parts on a material, such a voltage can cause a conductive leakage path across the surface of the material. The breakdown on the surface is called Tracking.

Material is classified into different categories depending on the CTI, being the lower CTI (and the IIIb category) the worst situation, with a higher risk of surface breakdown:

```
Category  →        ≤ IIIb <   ≤ IIIa <       ≤ II <      ≤ I
                 |--------------|--------------|--------------|-------->
   CTI  →    100           175           400           600
```

Figure 15- Comparative Tracking Index and corresponding Material Category

If we do not know our material CTI, category IIIb shall be assumed.

3.3 Calculation Procedure for a Food Mixer

Let us continue with the previous example on chapter 2, with reference to our household appliances and the IEC 60335-1 and IEC 60335-2-14 standards. In this case, we have only one table (table 17 of the standard).

Working Voltage V	Creepage distance [mm] Pollution Degree					
		2			3	
	1	Material Group			Material Group	
		I	II	IIIa/IIIb	I	II	IIIa/IIIb
≤ 50	0,18	0,6	0,85	1,2	1,5	1,7	1,9
125	0,28	0,75	1,05	1,5	1,9	2,1	2,4
250	0,46	1,25	1,8	2,5	3,2	3,6	4,0
400	1	2,0	2,8	4,0	5,0	5,6	6,3
500	1.3	2,5	3,6	5,0	6,3	7,1	8,0
630 < V ≤ 800	1.8	3,2	4,5	6,3	8	9	10
....	.....	.....	.....	.....	.....	.....	.....

Figure 16 - Table 17 of IEC 60335-1

For working voltage higher than 50V and lower/equal than 630 V, the value of the creepage distance may be found by interpolation.

Nonetheless, interpolation is rarely used among engineers, always preferring the higher value of the interval for simplicity sake and at the safety advantage, unless the space is such an issue, that the reduction of space gained with interpolation is valuable.

The standard also specifies that for pollution degree 3, the material group IIIb is allowed only if the working voltage does not exceed 50V.

For basic and supplementary insulation, we refer to Table 17 of IEC60335-1.

For reinforced insulation, while for clearance we were asked to refer to the subsequent value, here we were asked to double the distance.

As we mentioned above, with respect to clearance, here we need to know the material CTI, usually provided by the manufacturer, on the material datasheet.

In Figure 17, we can find the CTI of several materials available on the market.

Materials group	Tg	CTEz	ε_r(1 MHz/1 GHz/10 GHz)	Proof volt-age	Surface resistance	Conduc-tive Track-ing Index (CTI)	Water absorp-tion	Cu adhe-sion
	°C	ppm/K	-	KV/mm	MΩ	V	%	N/mm
Standard FR4	125°C-140°C	<70	4,7/4,3/-	50	10^7	>200	0,06	1,5
Modified FR4	135°C-180°C	<55	4,6/4,2/-	45	10^7	>200	0,06	1,5
FR4 halogen free	150°C-170°C	<40	5,0/4,8/4,6	50	10^8	>500	0,06	1,5
BT epoxy	Approx. 200°C	<40	4,4/4,1/-	70	10^8	>200	0,05	1,6
CE epoxy	Approx. 250°C	<25	3,9/3,7/3,5	65	10^7	>200	0,05	1,6
Polyimide	220°C-260°C	<55	4,0/3,8/3,8	45	10^8	>100	0,3	1
PTFE (pure)	200°C-230°C	<70	2,6/2,4/2,2	45	10^7	>600	0,04	1,3
RO3000	-	<40	3,0/2,8/2,6	30	10^7	>600	0,1	2,5
RO4000	Approx. 280°C	<45	3,3/3,0/2,8	30	10^9	>600	0,04	1,0

Figure 17 - Material datasheet

As a general rule, if from our calculations it emerges that creepage is lower than clearance, then we should adopt the clearance distance for both.

3.4 Numerical Calculation Example

Let us consider the food mixer again, this time for the US market.

- Power supply -->**110 V (USA)**

- PCB material --> **STANDARD FR4**

- Insulation working voltage --> **110 V**

- Whether the circuit is s primary or secondary -->
  PRIMARY

- Pollution degree (1 to 4) --> **3**

- The insulation category that it is meant to be implemented (e.g. basic, supplementary, reinforced) -->
  REINFORCED

- Altitude --> **NOT REQUIRED**

- Overvoltage category (I to IV) --> **II**

- If we have a quality control program --> **NOT REQUIRED**

- Material group category (I, II, IIIa, IIIb) from the CTI -->
  DEPENDS ON THE SUPPLIER

- Standard to apply --> **IEC 60335-1 and IEC 60335-2-14**

Let us assume that the material of choice for the PCB is Standard   FR4.

Materials group	Tg	CTEz	ε_r(1 MHz/1 GHz/10 GHz)	Proof volt-age	Surface resistance	Conduc-tive Track-ing Index (CTI)	Water absorp-tion	Cu adhe-sion
	°C	ppm/K	-	KV/mm	MΩ	V	%	N/mm
Standard FR4	125°C-140°C	<70	4,7/4,3/-	50	10^7	>200	0,06	1,5
Modified FR4	135°C-180°C	<55	4,6/4,2/-	45	10^7	>200	0,06	1,5

Figure 18 - Standard FR4 CTI

From the Table in **Figure 18**, we can see that the manufacturer guarantees a CTI higher than 200. So let us find out in Figure 19 what the material group is.

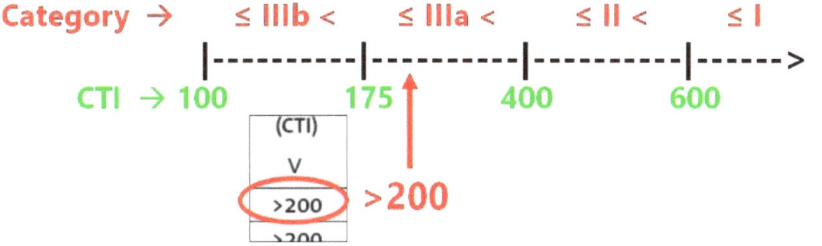

Figure 19 - Material category calculation

Therefore, we choose the worst value and the worst category we can expect given the manufacturer available data, so in this case CTI 200 and the corresponding Category IIIa.

Now we are able to continue with calculations.

We first select the Pollution degree (red frame in Figure 20), then the material group (green frame in the same Figure) and finally the working voltage (blue arrow and circle again Figure 20).

Working Voltage V	Creepage distance [mm] Pollution Degree						
	1	2 Material Group			3 Material Group		
		I	II	IIIa/IIIb	I	II	IIIa/IIIb
≤ 50	0,18	0,6	0,85	1,2	1,5	1,7	1,9
125	0,28	0,75	1,05	1,5	1,9	2,1	2,4
250	0,46	1,25	1,8	2,5	3,2	3,6	4,0
400	1	2,0	2,8	4,0	5,0	5,6	6,3

Pollution degree: 3

Working Voltage: 110

Materia Group: IIIa

Selected Value: 2.4 mm

Figure 20 - Calculation of the Creepage

According to the reinforced insulation rule, we take the value of the first column from the right, 2.4 mm, and then we double it.

Therefore, the creepage distance for our PCB is 4.8 mm, representing the reinforced insulation implementation (for what concern creepage distance) for a class II product.

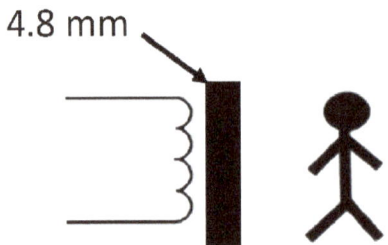

4.8 mm

Figure 21 - Class II with reinforced insulation

Chapter 4 – Solid insulation

4.1 Definition

We may also choose to interpose solid insulating material between two conductive parts or between a conductive part and an accessible surface.

In this case, we implement the so called solid insulation.

Solid insulation, like the other kind of techniques, needs to withstand the electrical stress that can be expected in the intended use (and foreseeable misuse).

Solid insulation relies on the characteristics of the material used as a barrier and is often adopted as supplementary insulation.

It needs to have a sufficient thickness and its minimum value is provided by different standards, depending on the final application intended use.

We can also use a barrier thinner than this minimum value, considering that the layers that form it have sufficient electric strength.

4.2 Calculation Procedure for Household Appliances

For household appliances, the minimum thickness value is dictated by IEC 60335-1:

- For a barrier implementing supplementary insulation: 1 mm

- For a barrier implementing reinforced insulation: 2 mm

If we use a multilayer material, each layer needs to withstand certain test values given in the following Table 7.

The supplementary insulation must count at least two layers, the reinforced one at least three.

Insulation	Test Voltage [V]			
	Rated voltage			Working voltage (U)
	SELV	≤ 150V	150 < V ≤ 250 (1)	>250
Basic insulation	500	1250	1250	1,2 U + 950
Supplementary insulation	-	1250	1750	1,2 U + 1450
Reinforced insulation	-	2500	3000	2,4 U + 2400

Figure 22 - Table 7 of IEC 60335-1

NOTE (1): For appliances with a rated voltage lower than 150 V, the column marked with (1) applies to parts with working voltage >150 and ≤ 250.

As we can see in Figure 22 in the first column from the right, the standard refers to working voltage. Indeed, even though we supply our appliance with a SELV voltage, we may have higher voltages inside, due to the intrinsic circuit functionalities.

This is quite common with switching mode power supplies, if we use charge pumps or in boost converters.

As an example, this diagram (courtesy of Texas Instruments) shows a circuit with such a characteristic.

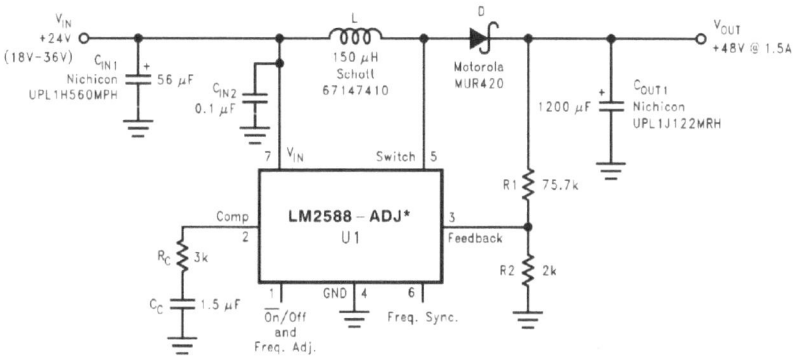

Figure 23 - Step-Up converter diagram.

The converter is fed with 24 VDC (SELV) and generates an output of 48 VDC (not SELV any longer).

4.3 Numerical Calculation Example

Let us continue with an example where we will select a proper enclosure that has to function as a secondary insulation at a rated voltage of 220 V.

If we choose an enclosure already available on the market, the manufacturer already provides us with information on the allowed rated voltage, like the one shown in Figure 24 from a real enclosure datasheet.

Protection class:	II - isolate protection acc. VDE 0106
Rated insulation voltage AC:	690 V
Rated insulation voltage DC:	1000 V

Figure 24 - Enclosure datasheet

In considering different kinds of enclosures, we will refer to information supplied by the manufacturer as in Figure 24.

We can see the insulation voltage (690 VAC and 1000 V DC) and the insulation class (Class II).

However, if we need a custom design, then it is important to consider the material features. In Figure 25 we have the information provided by a manufacturer with respect to different kinds of materials supplied by him.

AMORPHOUS THERMOPLASTICS		SEMICRYSTALLINE THERMOPLASTICS	
Dielectric strength - insulation (v/mil)		Dielectric strength - insulation (v/mil)	
• Ultem®	830	• Nylon (6 cast)	500-600
• PVC	544	• Acetal (Homopolymer)	500
• Kydex®	514	• Acetal (Copolymer)	500
• Noryl®	500	• PTFE	400-500
• Acrylic	430	• PEEK	480
• Polysulfone	425	• PPS	450
• PETG	410	• PET	400
• Polycarbonate	380	• PBT	400
• Radel R®	360	• Nylon (6/6 extruded)	300-400
		• PVDF (Kynar®)	280

Figure 25 - Dielectric strength

In our example, we want to design a Class II appliance with rated Voltage of 220 V and internal working voltages that does not exceed it.

We decide to implement a basic insulation with clearance/creepage plus a supplementary insulation using solid insulation.

We know that for supplementary insulation, the thickness shall be at least 1 mm, roughly 40 mils (1 mil = 0,0254 mm)

From Table 7 of IEC60335-1 shown in Figure 26, the solid barrier has to withstand 1750 V under test.

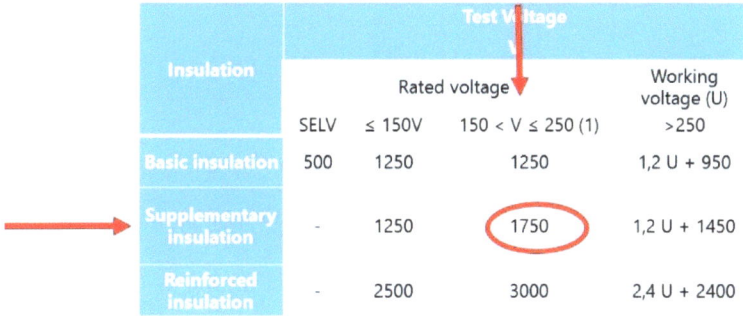

Insulation	Test Voltage			
	Rated voltage			Working voltage (U)
	SELV	≤ 150V	150 < V ≤ 250 (1)	>250
Basic insulation	500	1250	1250	1,2 U + 950
Supplementary insulation	-	1250	1750	1,2 U + 1450
Reinforced insulation	-	2500	3000	2,4 U + 2400

Figure 26 - Test voltage – Table 7 IEC 60335-1

So if we choose PET, from the above manufacturer, we have 1 mm = 39.37 mils, with 400V/mil --> It can withstand 15.748 V.

Otherwise, if we choose to perform a reinforced insulation with a solid insulation, we have a minimum thickness of 2 mm, 78.74 mils.

Again, with PET we would have 31,496 V, far greater than the 3,000 requested in the row of Table 7 (Figure 26) for reinforced insulation.

If the internally generated voltage is 280V, we should apply the fifth column formula 2,4 * 280 + 2400= 3072 V.

With these thicknesses in place, we may be quite sure that the test with the voltages in Table 7 will pass and the product will be compliant with these safety requirements.

Chapter 5 – Insulation diagrams

5.1 Definition and Scope

Insulation diagrams are graphic illustrations explaining how the insulation has been performed in our design.

They should not show more details than is required to illustrate the implemented barriers.

They are useful:

- Before the design: to choose and justify the insulation strategy

- For regulatory purposes: for instance, for medical devices it is required to have them in our technical folder or device master file, so to prove how we made it safe

- For design update: over time other engineers may work on the same project, these diagrams make them better understand our design

Being a graphical and representation tool, we resort to the abbreviation explained below:

- LOP: Level of Protection.

- Basic Insulation (BI): It is a spacing or a physical insulation barrier providing 1 LOP.

- Supplementary Insulation (SI): It is also a spacing or a physical insulation barrier providing 1 LOP.

- Double Insulation (DI): It is BI + SI and provides 2 LOP. Note that BI + BI does not result in DI.

- Reinforced Insulation (RI): It is a single spacing or physical insulation barrier that provides 2 LOP.

- Protective Impedance (PI): It is a component (such as a resistor) that provides 1 LOP.

- Protective Earth (PE): It is a well-grounded part that provides 1 LOP.

- Operational Protection (OP): It is an insulation that is not suitable as a means of protection, so it provides no LOP.

5.2 Examples

Let us examine a class I example supplied with 240 V and no secondary circuit, pollution degree 2, overvoltage category II and no altitude correction.

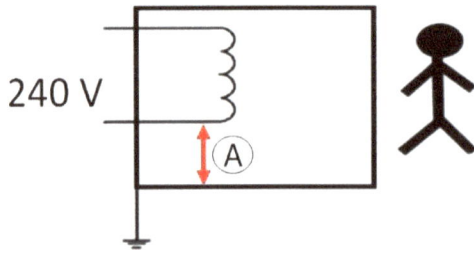

Figure 27 - Insulation diagram, Class I

In class I, one level of protection is provided by the protective earth connection, therefore we only need to take care of the basic insulation, for which we calculate clearance and creepage distances as explained in Chapter 2 and Chapter 3.

We may justify two level of protections with: 1 LOP = "A" implemented below, 1 LOP = "PE" protective earthing.

Then we will report everything in a Table in this way:

Insul. type	Max voltage V	Clearance		Creepage		Barrier Dielectric Strength			Standard used
		Required	Implemented	Required	Implemented	Required	Solid insulation material	Thickness	
A	BI	240	1.5 mm	1.5 mm	2.5 mm	3 mm	N/A	N/A	N/A	IEC 60335-1

Figure 28 - Insulation diagram Table

Therefore, we may provide documentary evidence on the kind of insulation, voltage, reference standard, required distance and the one actually being used.

As we can see, they have the ability to explain the insulation strategy clearly and with no ambiguity.

For Class II products, we would rather have a Diagram as in Fig. 29.

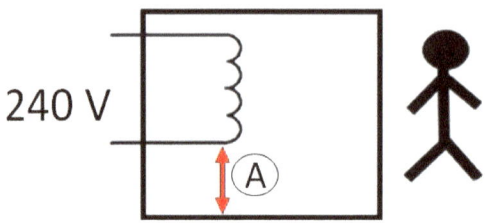

Figure 29 - Insulation diagram, Class II

In this case "A" implements 2 LOP and the corresponding Table is:

Insul. type	Max voltage V	Clearance		Creepage		Barrier Electric Strength			Standard used
		Requir ed	Implem ented	Requir ed	Implem ented	Requir ed	Solid insulation material	Thickness	
BI	240	1.5 mm	1.5 mm	2.5 mm	2.5 mm	N/A	N/A	N/A	IEC 60335-1
SI	240	1.5 mm	1.5 mm	2.5 mm	2.5 mm	N/A	N/A	N/A	IEC 60335-1

Figure 30 - Insulation Diagram Table

If we choose a solid insulation for the supplementary protection, we have a Table like the following one:

Insul. type	Max voltage V	Clearance		Creepage		Barrier Electric Strength			Standard
		Requir ed	Implem ented	Requi red	Implem ented	Requir ed	Solid insulation material	Thickness	
BI	240	1.5 mm	1.5 mm	2.5 mm	2.5 mm	N/A	N/A	N/A	IEC 60335-1
SI	240	N/A	N/A	N/A	N/A	2500 V	PET	2 mm	IEC 60335-1

Figure 31 - Insulation Diagram Table

Finally, for reinforced insulation we have:

Insul. type	Max voltage V	Clearance		Creepage		Barrier Electric Strength			Standard
		Requir ed	Implem ented	Requir ed	Implem ented	Requir ed	Solid insulation material	Thickness	
A	RI	240	3 mm	3 mm	5 mm	5 mm	N/A	N/A	N/A	IEC 60335-1

Figure 32 - Insulation Diagram Table

In some cases, insulation diagrams can become rather complex, especially if we develop a medical application according to IEC 60601-1.

Nonetheless, once the basic principles have been understood, it is just a matter of application analysis, by using the appropriate standard for distance and thickness calculations and take time to find and report in a table all the relevant solutions and insulation layers.

Chapter 6 – Conclusion

First of all, thank you for reading this book.

In this book we have learnt basic definitions and how to use a food mixer as an example, we have seen how to calculate creepage, clearance, solid insulation and how to draw and use insulation diagrams.

We have mixed basic concepts and definitions in a practical approach by using international standards. For every application, we can find a proper standard that guides us with the state of the art on safety matters, including electrical safety.

It is very important that before starting a project you check their availability (e.g. at standardization agencies websites www.iso.org www.iec.ch) and follow them thoroughly.

The main aim of this work has been to provide a quick guide to concepts often believed more complex and mysterious than what they actually are and we hope this has demystified most of the clouds that are around a topic often made more theoretical and complex than what it really is.

About the Author

Marco Catanossi has a MSc in electronics engineering and he is an expert in product safety and certifications, helping many companies from different countries to successfully put their product on the market. The Author has also experience as assessors on product safety on behalf of the Court of Justice and he is member of the European Union engineering panel for innovation projects evaluation. Since 2008, Marco has built a consulting business helping manufacturers to meet regulatory and technical requirements worldwide. He is a dedicated professional with a high specialization in electrical and electronics products safety, a deep insight in product liability and a strong practical approach to product engineering and design.

www.ingramcontent.com/pod-product-compliance
Lightning Source LLC
Chambersburg PA
CBHW040927180526
45159CB00002BA/640